The Sun

by Christina Leaf
Illustrated by Natalya Karpova

BELLWETHER MEDIA
MINNEAPOLIS, MN

Blastoff! Missions takes you on a learning adventure! Colorful illustrations and exciting narratives highlight cool facts about our world and beyond. Read the mission goals and follow the narrative to gain knowledge, build reading skills, and have fun!

Traditional Nonfiction

Narrative Nonfiction

Blastoff! Universe

MISSION GOALS

> FIND YOUR SIGHT WORDS IN THE BOOK.

> LEARN ABOUT THE SUN.

> FIND SOMETHING IN THE BOOK THAT YOU WOULD LIKE TO LEARN MORE ABOUT.

This edition first published in 2023 by Bellwether Media, Inc.

No part of this publication may be reproduced in whole or in part without written permission of the publisher. For information regarding permission, write to Bellwether Media, Inc., Attention: Permissions Department, 6012 Blue Circle Drive, Minnetonka, MN 55343.

Library of Congress Cataloging-in-Publication Data

Names: Leaf, Christina, author.
Title: The sun / by Christina Leaf.
Description: Minneapolis, MN : Bellwether Media, 2023. | Series: Blastoff! missions. Journey into space | Includes bibliographical references and index. | Audience: Ages 5-8 | Audience: Grades 2-3 | Summary: "Vibrant illustrations accompany information about the Sun. The narrative nonfiction text is intended for students in kindergarten through third grade"-- Provided by publisher.
Identifiers: LCCN 2022006863 (print) | LCCN 2022006864 (ebook) | ISBN 9781644876589 (library binding) | ISBN 9781648348426 (paperback) | ISBN 9781648347047 (ebook)
Subjects: LCSH: Sun--Juvenile literature.
Classification: LCC QB521.5 .L43 2023 (print) | LCC QB521.5 (ebook) | DDC 523.7--dc23/eng20220422
LC record available at https://lccn.loc.gov/2022006863
LC ebook record available at https://lccn.loc.gov/2022006864

Text copyright © 2023 by Bellwether Media, Inc. BLASTOFF! MISSIONS and associated logos are trademarks and/or registered trademarks of Bellwether Media, Inc.

Editor: Betsy Rathburn   Designer: Jeffrey Kollock

Printed in the United States of America, North Mankato, MN.

This is **Blastoff Jimmy**! He is here to help you on your mission and share fun facts along the way!

Table of Contents

A Solar Eclipse	4
The Center of the Solar System	8
An Active Star	14
Glossary	22
To Learn More	23
Beyond the Mission	24
Index	24

A Solar Eclipse

solar eclipse

You are ready! You gaze up at the sky with special glasses. A total **solar eclipse** is starting! The sky gets dark and the air becomes chilly.

You watch as the Moon fully covers the Sun. You can only see the Sun's **corona**.

Hmmm...

The Center of the Solar System

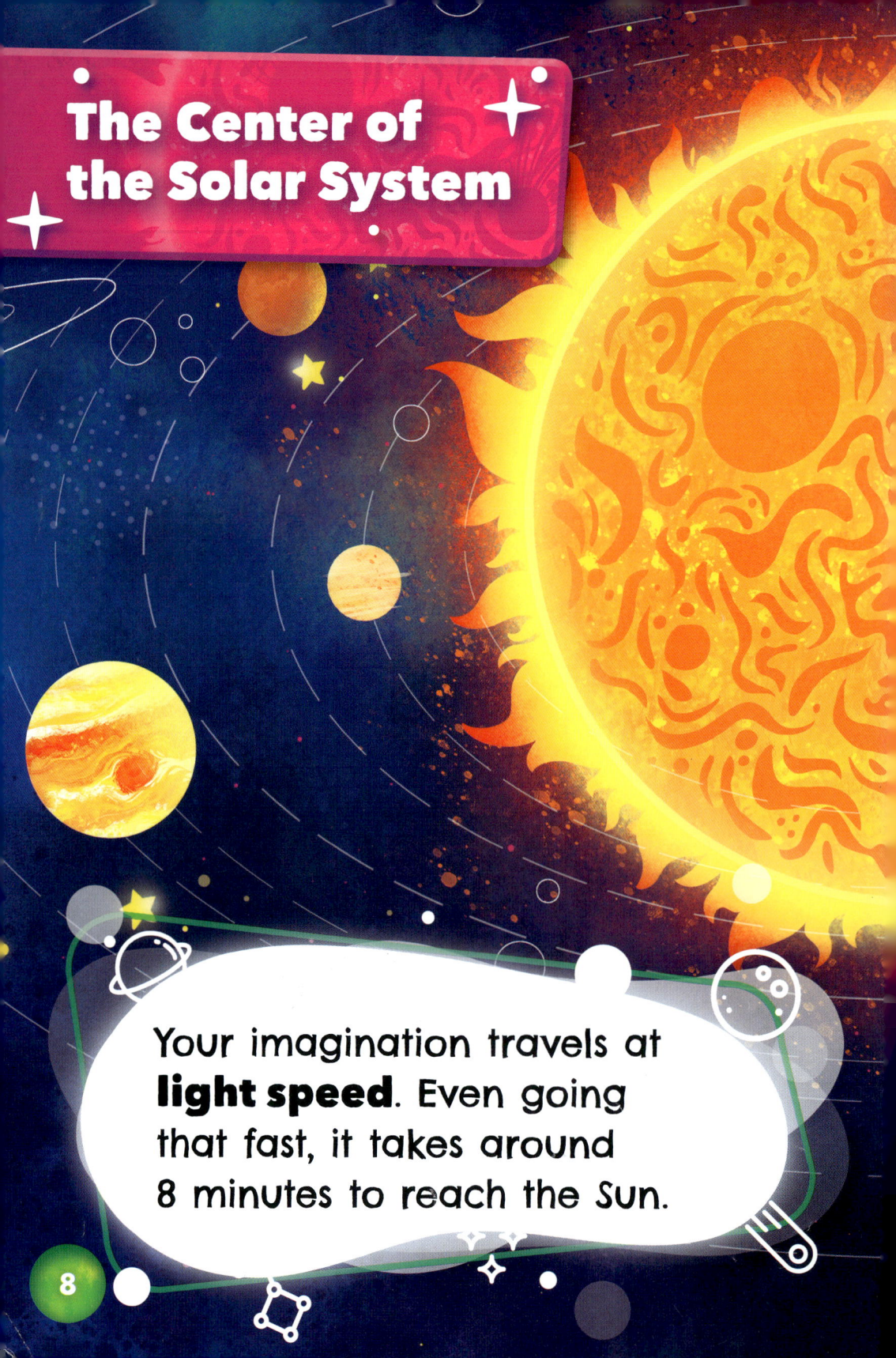

Your imagination travels at **light speed**. Even going that fast, it takes around 8 minutes to reach the Sun.

The Sun is at the center of the **solar system**.
It is about 93 million miles (150 million kilometers) from Earth.

You are in awe of the Sun's huge size. More than 1 million Earths could fit inside. It feels silly to call it a **dwarf star**!

JIMMY SAYS
The Sun is a yellow dwarf. This kind of star is medium-sized.

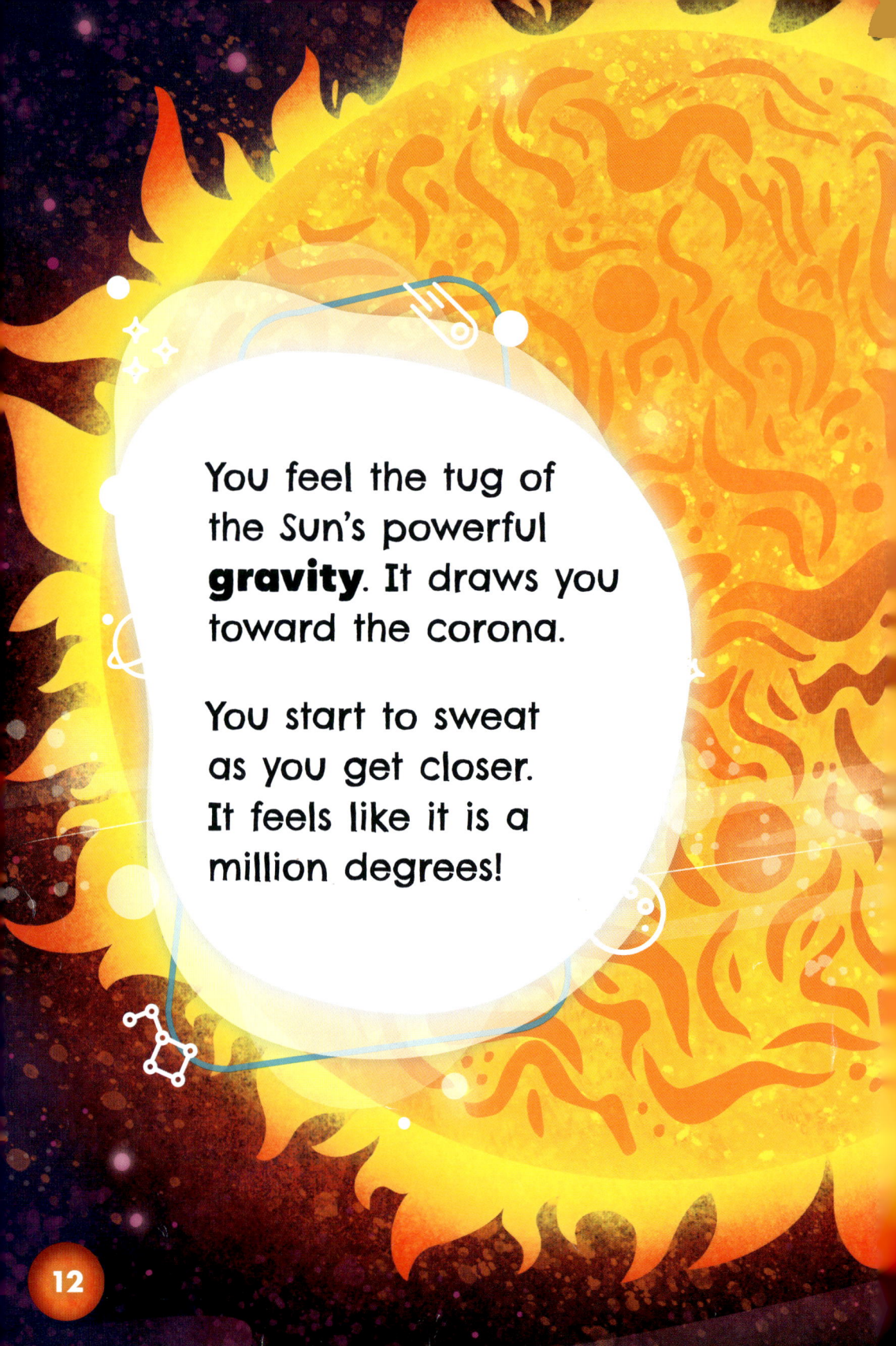

You feel the tug of the Sun's powerful **gravity**. It draws you toward the corona.

You start to sweat as you get closer. It feels like it is a million degrees!

An Active Star

You can see the Sun's surface moving. It must be because of the **nuclear reactions** in the **core**. These make the Sun's heat and light!

You move closer. There is a dark spot on the Sun. Is that a **sunspot**? You feel slightly cooler as you move in front of it.

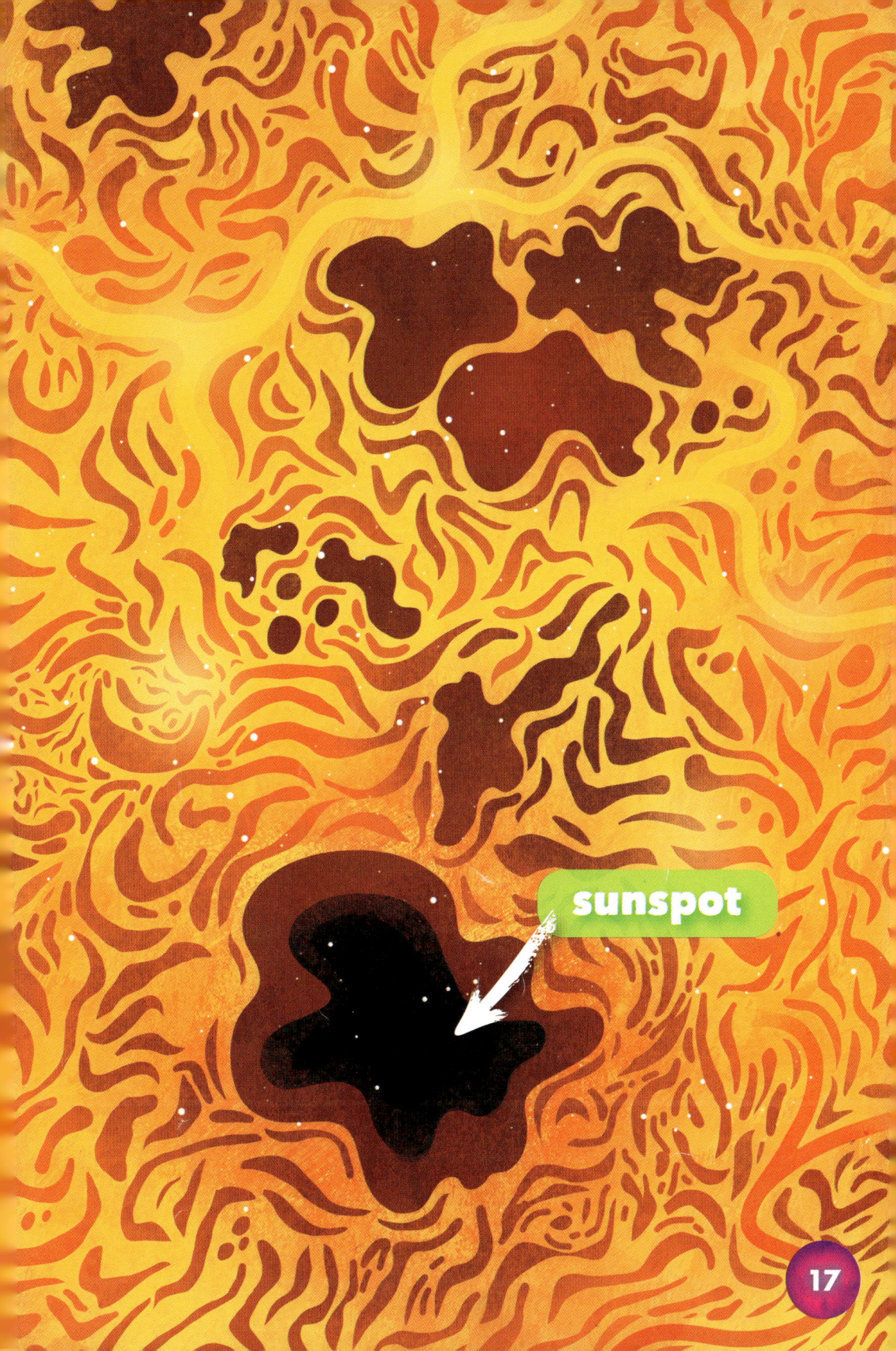

solar flare

Yikes! A **solar flare** started near the sunspot. You barely escape getting zapped. Time to go home!

Back home, the Sun is uncovered again. The eclipse is over. Everything is warmer and lighter. You had a good visit. But you like the Sun better at a distance!

Glossary

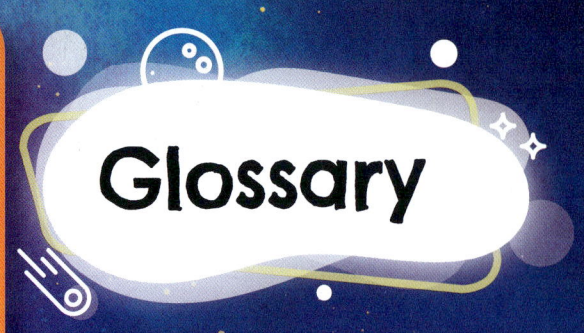

core—the center of the Sun

corona—the outermost part of the Sun

dwarf star—a kind of star that is fairly dim and smaller than other kinds of stars; the Sun is a dwarf star.

gravity—a force that pulls objects toward each other

light speed—the speed at which light travels; nothing travels faster than light.

nuclear reactions—events between materials that give off huge amounts of energy

solar eclipse—an event in which the Moon moves between the Sun and Earth; a total solar eclipse happens when the Moon covers the entire Sun.

solar flare—an outburst of energy from the surface of the Sun

solar system—the group of planets, moons, asteroids, and other bodies that circle around the Sun

sunspot—a dark spot on the surface of the Sun that is cooler than its surroundings

To Learn More

AT THE LIBRARY

Betts, Bruce. *Super Cool Space Facts: A Fun, Fact-filled Space Book for Kids.* Emeryville, Calif.: Rockridge Press, 2019.

Leaf, Christina. *The Inner Planets.* Minneapolis, Minn.: Bellwether Media, 2023.

McAnulty, Stacy. *Sun! One in a Billion.* New York, N.Y.: Henry Holt and Company, 2018.

ON THE WEB

FACTSURFER

Factsurfer.com gives you a safe, fun way to find more information.

1. Go to www.factsurfer.com.

2. Enter "Sun" into the search box and click .

3. Select your book cover to see a list of related content.

BEYOND THE MISSION

> WHAT IS ONE NEW FACT YOU LEARNED FROM THE BOOK?

> WOULD YOU LIKE TO VISIT THE SUN? WHY OR WHY NOT?

> WHAT FEATURES WOULD YOU ADD TO YOUR SPACESHIP IF YOU VISITED THE SUN?

Index

core, 14
corona, 6, 7, 12, 13
distance, 9
dwarf star, 10, 11
Earth, 7, 9, 10
gas, 15
gravity, 12
heat, 14, 20
light, 14, 20
light speed, 8
Moon, 6
nuclear reactions, 14
size, 10, 11
solar eclipse, 5, 6, 7, 20

solar flare, 19
solar system, 9
sunspot, 16, 17, 19
surface, 14, 15
temperature, 12, 13, 16